SCIENTISTS WHO STUDY
WILD ANIMALS

Mel Higginson

The Rourke Corporation, Inc.
Vero Beach, Florida 32964

Edited by Sandra A. Robinson

PHOTO CREDITS
© Mel Higginson: cover, title page, pages 10, 12, 13, 15, 18, 21;
courtesy Idaho Department of Parks & Recreation: page 4;
© Barry Gordon: pages 7 and 17; courtesy of Connecticut
Department of Environmental Protection: page 8

Library of Congress Cataloging-in-Publication Data

Higginson, Mel, 1942-
   Scientists who study wild animals / by Mel Higginson.
   p. cm. — (Scientists)
   Includes index.
   ISBN 0-86593-374-X
   1. Zoologists—Juvenile literature. 2. Zoology—Vocational
guidance—Juvenile literature. [1. Zoologists. 2. Zoology—
Vocational guidance. 3. Occupations. 4. Vocational guidance.]
I. Title. II. Series: Higginson, Mel, 1942-   Scientists.
QL50.5.H54   1994
591'.092—dc20   94-6997
                                       CIP

Printed in the USA   AC

TABLE OF CONTENTS

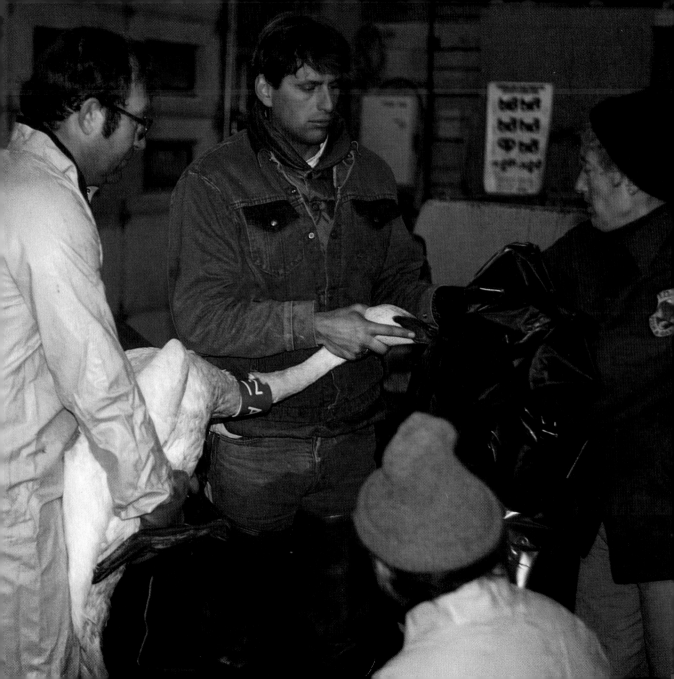

SCIENTISTS WHO STUDY WILD ANIMALS

Thanks to wildlife **biologists,** the scientists who study wild animals, the secret lives of animals are now less secret.

By learning more about wild animals, scientists can find better ways to protect them. The work of wildlife scientists has helped save North American animals such as the whooping crane, bald eagle and grizzly bear.

Wildlife biologists discovered more secrets of trumpeter swans after they began using neckbands on the birds

WHAT WILDLIFE SCIENTISTS DO

Wildlife scientists learn as much about a **species,** or kind, of animal as they can. They find out how an animal spends its time. They find out where and when the animal travels. They learn about what an animal eats — and what eats it.

Scientists cannot always watch wild animals. They put radio collars on some animals. The collars send a signal that scientists can receive on instruments in a truck, airplane or satellite.

A scientist prepares a radio collar for a drugged grizzly in Montana

KINDS OF WILDLIFE SCIENTISTS

Wildlife scientists are trained to study only certain groups of animals. For example, scientists may study snakes, alligators and other reptiles. Some scientists study mammals — seals, bears, lions and other furry creatures. Other wildlife scientists are experts on the lives of birds or fish.

Wildlife biologists often try to find out why certain animals have become **endangered,** or rare. What the scientists learn can help save the animals from becoming **extinct.**

A snorkeling scientist studies fish and other life in a Connecticut stream

WHERE WILDLIFE SCIENTISTS WORK

Wildlife scientists work and camp in the wild, rugged places where some wild animals live. Scientists climb after bighorn sheep and cougars in Western mountains. They track wolves in Northern forests, and study snow geese on the Arctic **tundra.**

Scientists working at night wade into saltwater lakes. There they find crocodiles.

Wildlife study is not all outdoor work, however. Scientists sometimes study captive animals. They also work in **laboratories** and offices.

Working at night in the Florida Keys, wildlife scientists caught this young crocodile, which was measured, tagged and released

Wildlife scientist Paul Moler prowls the north Florida wetland where he discovered a new species of frog by hearing its call

Wildlife scientists at the International Crane Foundation in Baraboo, Wisconsin, help insure the future of whooping cranes and other rare cranes

THE IMPORTANCE OF WILDLIFE SCIENTISTS

The work of wildlife scientists helps people understand the needs of wild animals.

The scientists' work also helps the animals directly. For example, wildlife scientists use photos taken from airplanes to count geese. The count helps them decide how many geese hunters should be allowed to shoot.

Lawmakers use information from wildlife biologists to decide whether an animal needs help. America's Endangered Species Act gives special protection to animals that are in danger of becoming extinct.

Wildlife scientists and the Endangered Species Act have helped the bald eagle population grow

TRACKING THE GRIZZLY

Wildlife scientists often study endangered animals — pandas, tigers, crocodiles and hundreds of other species. A few scientists in western North America study the endangered grizzly bear.

Scientists capture grizzlies by putting meat in a bear trap. A scientist uses drugs to put the trapped bear to sleep, and checks its health. Then the scientist fits the bear with a radio collar. The bear soon wakes up and runs off. Now the scientist can track the grizzly.

By learning more about the grizzly's habits, scientists hope to save it from extinction.

Wildlife scientists work quickly to collar and check the health of a grizzly before the drug wears off

DISCOVERIES OF WILDLIFE SCIENTISTS

Scientists have discovered many ways to help save wild animals. Scientists learned that whooping cranes usually lay two eggs, but that they usually raise just one baby.

Wildlife scientists began to "steal" one egg from each nest of the endangered whoopers. They placed the "stolen" eggs in the nests of common sandhill cranes. The sandhills raised the young whoopers, helping the tiny population of whooping cranes to grow.

Experiments with the nests and eggs of sandhill cranes helped scientists plan the release of endangered whooping cranes in Florida

LEARNING TO BE A WILDLIFE SCIENTIST

Wildlife scientists are deeply interested in nature and the outdoors. Almost all of them learned early in their lives about wild creatures.

Later, wildlife scientists studied plants and animals in high school and college.

Wildlife scientists have at least four years of college. Many continue college for another two or three years, so they can carefully study one kind of animal.

This wildlife science student is putting a small metal tag on one foot of each snow goose gosling in the nest

CAREERS FOR WILDLIFE SCIENTISTS

Wildlife scientists work for zoos, universities, governments and some private groups or businesses. A scientist may be hired to study a single kind of animal, or species of animal. A scientist may also be hired to study many kinds of animals that live in a certain area.

Sometimes the scientist decides what he or she wants to study. If it is a good idea, the government may pay the scientist to do the work.

Glossary

biologist (bi AHL uh gist) — a scientist who studies living things

endangered (en DANE jerd) — in danger of no longer existing; very rare

extinct (ex TINKT) — no longer existing

laboratory (LAB rah tor ee) — a place where scientists can experiment and test their ideas

species (SPEE sheez) — a certain kind of animal within a closely related group; for example, a *whooping* crane

tundra (TUN druh) — the treeless "carpet" of low-lying plants in the Far North and on mountains above the tree line

INDEX